Exploration of the Asteroid Belt, a new approach.

Patrick H. Stakem

(c) 2018

Number 26 in the Space Series

Table of Contents

Table of Contents

Introduction..3
Author...3
    Challenges of the Asteroid Belt...............................4
Enabling Technologies...12
    Overall Architecture..15
    Batteries and solar panels...17
    Cubesats..17
    Dispenser/Mothership...17
    Cubesat Swarm...19
    Avionics Suite...21
    Compute cluster of convenience 22
    Surface Lander...23
    Communications..25
    Communication methodology within the Cubesat swarm. ..27
    Rad-Hard Software 29
    Onboard databases..31
Wrap-Up..34
Glossary of Terms...35
Bibliography...42
    ANTS/SWARM references......................................49
Resources..54
If you enjoyed this book, you might also be interested in some of these..56

Introduction

This book discusses the exploration of the Asteroid Belt. This is very difficult, because of the millions of diverse targets to visit. The sizes range from dust particles to the dwarf planet Ceres. The area of the asteroid belt was searched by astronomers in the 18th century, based on the lack of a planet at a predicted orbit by the Titus Bode Law.

Before the 1890's, astronomy depended on human observation through the telescope, and this was difficult with small objects, as compared to planets. In 1891, astro-photography began producing better results.

This book will provide an overview and description of asteroid missions. It will then suggest an approach to exploring the asteroid belt is with a large swarm of small spacecraft.

Author

Mr. Patrick H. Stakem has been fascinated by the space program since the Vanguard launches in 1957. He received a Bachelors degree in Electrical Engineering from Carnegie-Mellon University, and Masters Degrees in Physics and Computer Science from the Johns Hopkins University. At Carnegie, he worked with a group of undergraduate students to re-assemble, modify, and operate a surplus missile guidance computer, which he was later able to get donated to the Smithsonian.

He began his career in Aerospace with Fairchild Industries on the ATS-6(Applications Technology Satellite-6) program, a communication satellite that developed much of the technology for the TDRSS (Tracking and Data Relay Satellite System). He followed the ATS-6 Program through its operational phase, and worked on other projects at NASA's Goddard Space Flight Center including the Hubble Space Telescope, the International Ultraviolet Explorer (IUE), the Solar Maximum Mission (SMM), FUSE, some of the Landsat missions, and Shuttle. He was posted to NASA's Jet Propulsion Laboratory for the Mars-Jupiter-Saturn (MJS-77), Mission which later became the *Voyager* mission, and is still operating and returning data from outside the solar system at this writing. He initiated and lead the international Flight Linux Project for NASA's Earth Sciences Technology Office. He is the recipient of a Shuttle Program Manager's Commendation Award, and has completed 42 NASA Certification courses. He has two NASA Group Achievement Awards, and the Apollo-Soyuz Test Program Award.

Mr. Stakem has been affiliated with the Whiting School of Engineering of the Johns Hopkins University, The graduate Computer Science Dept. of Loyola University in Maryland, and Capitol Institute of Technology. Mr. Stakem supported the Summer Engineering Bootcamp Project at Goddard Space Flight Center.

Challenges of the Asteroid Belt

There are millions of asteroids, mostly in the inner solar

system. The main asteroid belt is between Mars and Jupiter. Each asteroid may be unique, and some may provide needed raw materials for Earth's use. There are three main classifications: carbon-rich, stony, and metallic.

The physical composition of asteroids is varied and poorly understood. Ceres, the Dwarf planet in the Asteroid belt, appears to be composed of a rocky core covered by an icy mantle, whereas Vesta may have a nickel-iron core. Hygiea appears to have a uniformly primitive composition of carbonaceous chondrite. Somewhat smaller than Ceres are Vesta and Passas. Many of the smaller asteroids are piles of rubble held loosely together by gravity. Some have moons themselves, or are co-orbiting binary asteroids. The asteroids rotate in their orbits, much like the planets. The bottom line is, asteroids are numerous and diverse.

It has been suggested that asteroids might be used as a source of materials that may be rare or exhausted on earth (asteroid mining) or materials for constructing space habitats or as refuelling stations for outer planet missions. Materials that are heavy and expensive to launch from Earth may someday be mined from asteroids and used directly for space manufacturing.

The asteroids are not uniformly distributed. In the asteroid belt, the Kirkwood gaps are relatively empty spots. This is caused by orbital resonance of the asteroids with Jupiter.

Asteroids vary from 1000 km in diameter to 1 meter and less. The larger ones are roughly spherical, while the mid-size and smaller ones are irregular in shape. Dactyl is the moon of the asteroid Ida.

Later we will describe an approach to exploration of the asteroid belt, and discusses the advantages and challenges of using a Swarm approach. This approach was first explored as a multi-student study of an alternative to the Juno Mission to Jupiter. We worked within the mission's size, weight, and power envelopes. Staying within the parameters of the mission, we were able to accommodate 1,000-1U Cubesats. We chose 333 3-U, and explored issues in local control, networking communications, and implementation in Open Source. This became the baseline for defining the Asteroid Swarm Mission. The Mission Parameters are quite different.

Exploring the asteroids requires a diverse and agile system. A swarm of small spacecraft with different capabilities can be used, combining into Teams to address situations and issues discovered in situ. A intelligent swarm solution to resource exploration of the asteroid belt was proposed by Curtis, et al, in 2003. The concept of Cubesats was not advanced enough at the time for the authors to specifically mention them, but they explored issues in local control, networking communications, and implementation in Open Source.

The Asteroid belt is located between 2.0 AU and 3.2 AU,

with Mars being at 1.67 AU. The belt is about 1 AU thick. It forms a continuous, sparsely populated ring around the Sun. The average distance between the objects in the belt is large, and solid objects are sparse. The volume of the belt is about 1.7×10^9 cubic miles.

A driver in asteroid belt exploration is the shear number of things that need to be explored. We can no longer afford to send dedicated single spacecraft missions to all the potential targets in our solar system. Although there are fewer than 10 planets, and less than 200 moons, there are millions of asteroids, mostly in the inner solar system. Each asteroid may be unique. The asteroid belt contains some 750,000 rocks larger than one kilometre in diameter.

The radiation environment of the asteroid belt is mostly of solar origin, with some Galactic cosmic rays. There is no overall magnetic field to deflect energetic particles. A lot of the material in the belt is dust sized, and will pose a damage threat to spacecraft.

Asteroid 4-Vesta may have a nickel-iron core. It was explored by the Dawn spacecraft in 2012. DAWN showed that Ceres is an Ocean world, and thus of interest to exo-biologists. 10-Hygiea appears to have a uniformly primitive composition of carbonaceous chondrite. Many of the smaller asteroids are piles of rubble held together loosely by gravity. The asteroids with moons are described as "rubble piles."The bottom line is, asteroids are numerous and diverse.

Orbiting irregularly shaped bodies is challenging, due to the irregular gravity field. This makes station keeping and attitude control a much bigger problem.

Extra-solar asteroids have been observed orbiting other stars.

NASA's Infrared Astronomical Satellite (IRAS) discovered three new asteroids. The Wide Field Infrared Survey Explorer (WISE) launched in 2009. It completed observations of over 33,000 new asteroids between 2009 and 2011. the Galileo mission was the first to fly by asteroids, on its way to Jupiter. The first asteroid studied from orbit was 433 Eros, by the NEAR Shoemaker spacecraft. The Dawn mission is currently orbiting the dwarf planet in the belt, Ceres. It has used up all of its fuel, at this point.

The asteroid Ryugu was discovered in 1999. The Japanese mission Hayabusa-2 launched in 2014, and arrived at the asteroid in 2018. It will attempt a sample return mission in late 2020. The sample is currently scheduled to be taken in early 2019, with an Earth return in 2020. The mission includes four surface rovers. It's crowded around Ryugu, as a German-French joint project MASCOT (Mobile Asteroid Surface Scout) arrived 10 days after the Japanese mission, but only had a 16-hour operational life.

Asteroid observation data has been provided by the Herschel Space Observatory and the Wide-field Infrared

Survey Explorer (WISE), among others. After we began to launch telescopes to Earth orbit, such as HST, we began to see much more detail. The NEAR spacecraft was the first dedicated asteroid mission, it orbited, then landed on Eros. The OSIRIS-Rex sample return mission is on a mission to the asteroid 101955 Bennu (the author worked on this mission). It arrived in December of 2018, at 19 km. stand-off. It will do a landing, and a sample will be taken. This is slated to return to Earth by September of 2023. That's one asteroid, one sample.

Russia and China have also had asteroid missions. They key question about the asteroid belt is, based on how little we know, is this representative of the entire population. Best guess is, no, there are surprises awaiting.

Planetoid Mines, a private mining corporation in New Mexico, has joined one of NASA's Centennial Challanges.

You may not have known that asteroids can have moons of their own. Well, Dimorphos, an asteroid moon-let of Didymos, was deliberately smacked by a spacecraft called DART, the double asteroid redirection test. The impact was observed, and its result was a reduction of 10 minutes of its orbital time. The crash was filmed.

Bottom line, we can do it. We can alter the orbit of a potentially hazardous asteroid, if need be.

Given the size of the belt, and the relatively sparse distribution of the asteroids, we could spend hundreds of years exploring every one. The rest of this book discusses an approach to using a distributed swarm of small spacecraft to visit more targets, and to co-operate in observations. Critical is the mothership, which is the transport vehicle, the communications relay with Earth, and manages the swarm taking date, during periods when communications with the Earth is not possible.

In 2003, NASA was looking at alternatives for exploration of the asteroid belt. One of these was PAM – Prospecting Asteroid Mission. It involved 1,000 spacecraft in a swarm to travel to the asteroid belt. They would have a common architecture, with 10 sub-swarms of 100 spacecraft each. They would mass about 1 kg. each. They would address targets of opportunity on site. They were to travel to the asteroid belt using solar sail technology. Their primary purpose was to search for resources and material with astro-biologically relevant signatures.

This would rely heavily on the ANTS mission architecture concept, which was itself predated by the BEES. BEES was the Biologically Inspired Engineering for Exploration Systems.

NASA did a lot of work in the 1970's and 1980's on ANTS – the Autonomous Nano-Technology Swarm. The Basic concepts were defined, but the implementation was not yet ready. Now, Cubesat-type architectures can address that. The overall concept is many small units working together, self-configuring, self-healing, making in-situ decisions. No direct human control, not pre-programmed, but agile, making decision on the spot. A reference mission to the Asteroid Belt in 2020 was defined. Well, here we are, where's the mission? (This is partially addressed in the authors book, "A Cubesat Swarm Approach for Exploration of the Asteroid Belt, originally a student project that got to a high TRL). Another feature the swarm exhibits is teaming, social interaction between individual units. ANTS was hard to implement with the hardware and software available in the 1980's. It fits well into a Cubesat implementation.

In Swarm robotics, the key issues are communication between units, and cooperative behavior. The capability of individual units does not much matter; it is the strength in numbers. Ants and other social insects such as termites, wasps, and bees, birds, fish, and reindeer are models for swarm behavior. Self-organizing behavior emerges from decentralized systems that interact with members of the group, and the environment. Swarm intelligence is an emerging field, and swarm robotics is in its infancy. Co-operative behavior, enabled by software and intra-unit communications has been demonstrated. Software Agent technology is the key to implementation of the autonomous swarm.

We can postulate target selection according to mission goals, but also mention that mission goals change as data is collected at the site.

At the time, Cubesats as small, modular satellites, had not yet been defined. The BEES mission was seen as planetary surface exploration. There was also, LARA – ANTS-based Lander Amorphous Rover Antenna (Mission Concept).

For ground based mobility, a tetrahedral structure for the rover was considered and tested.

Enabling Technologies

This section will discuss some of the enabling technologies that will be used in the Cubesat Swarm Asteroid mission definition. Use of existing technologies from previous missions was a goal. An ops concept will be presented later. First, we have to review the concept of the technology readiness level (TRL).

The Technology readiness level (TRL) is a measure of a device's maturity for use. There are different TRL definitions by different agencies (NASA, DoD, ESA, FAA, DOE, etc). TRL are based on a scale from 1 to 9 with 9 being the most mature technology. The use of TRLs enables consistent, uniform, discussions of technical maturity across different types of technology. We will discuss the NASA's version here, which was the original definition from the 1980's.

Technology readiness levels in the National Aeronautics and Space Administration (NASA)

1. Basic principles observed and reported
This is the lowest "level" of technology maturation. At

this level, scientific research begins to be translated into applied research and development.

2. Technology concept and/or application formulated

Once basic physical principles are observed, then at the next level of maturation, practical applications of those characteristics can be 'invented' or identified. At this level, the application is still speculative: there is not experimental proof or detailed analysis to support the conjecture.

3. Analytical and experimental critical function and/or characteristic proof of concept.

At this step in the maturation process, active research and development (R&D) is initiated. This must include both analytical studies to set the technology into an appropriate context and laboratory-based studies to physically validate that the analytical predictions are correct. These studies and experiments should constitute "proof-of-concept" validation of the applications and concepts formulated at TRL 2.

4. Component and/or breadboard validation in laboratory environment.

Following successful "proof-of-concept" work, basic technological elements must be integrated to establish that the "pieces" will work together to achieve concept-enabling levels of performance for a component and/or breadboard. This validation must be devised to support the concept that was formulated earlier, and should also be consistent with the requirements of potential system applications. The validation is "low-fidelity" compared to the eventual system: it could be composed of ad hoc discrete components in a laboratory

TRL's can be applied to hardware or software, components, boxes, subsystems, or systems. Ultimately, we want the TRL level for the entire systems to be consistent with our flight requirements. Some components may have higher levels than needed.

5. Component and/or breadboard validation in relevant environment.

At this level, the fidelity of the component and/or breadboard being tested has to increase significantly. The basic technological elements must be integrated with reasonably realistic supporting elements so that the total applications (component-level, sub-system level, or system-level) can be tested in a 'simulated' or somewhat realistic environment.

6. System/subsystem model or prototype demonstration in a relevant environment (ground or space).

A major step in the level of fidelity of the technology demonstration follows the completion of TRL 5. At TRL 6, a representative model or prototype system or system - which would go well beyond ad hoc, 'patch-cord' or discrete component level breadboarding - would be tested in a relevant environment. At this level, if the only 'relevant environment' is the environment of space, then the model/prototype must be demonstrated in space.

7. System prototype demonstration in a space environment.

TRL 7 is a significant step beyond TRL 6, requiring an actual system prototype demonstration in a space environment. The prototype should be near or at the scale of the planned operational system and the demonstration must take place in space.

The TRL assessment allows us to consider the readiness and risk of our technology elements, and of the system.

Overall Architecture

In this concept, the Cubesats are the primary payload. The Mothership can be thought of as a very large Cubesat. The architecture is kept as close as possible.

Use of a common hardware bus and software architecture for all swarm members, to the greatest extent possible, is a goal. Only the sensor sets will be unique. A Cubesat model for the hardware, and NASA GSFC's Core Flight Software is baselined. A standard linux software operating environment and database will be used.

Each member of the swarm will be aware of other swarm members in close proximity. This will be facilitated by having the Mothership as the center of the coordinate system. It will determine its position by celestial navigation. The Cubesats will use the mothership their reference. The mothership will maintain, as part of its onboard database, the location of swarm members. It will also monitor for pending collisions and warn the participants. There will be rules concerning how close swarm members can get to each other, a virtual zone of exclusion. All Earth-based interaction with the swarm will be through the Mothership. Due to varying communication delays, tele-operation of the swarm from Earth is not feasible. The Swarm could be on the opposite side of the Sun from the Earth for extended periods. This is addressed by building autonomy into the system, and a large amount of non-volatile storage will be included for science data.

Each swarm member will be equipped with one or more cameras, not only for target investigation, but also for observing the position and relative motions of other swarm members.

Using standard linux clustering software (Beowulf), the Mothership and undeployed swarm members will be able to form an ad-hoc cluster computer to process science data in-situ. Within the Mothership, a LAN-based Mesh network software will be used. The Mothership's main computer will be a Raspberry-Pi based compute-cluster.

Complexity in a system generally derives from two parameters, the number of units, and the number of interactions. A swarm of Cubesats is complex, compared to a single spacecraft. This is balanced by the relative simplicity of the individual units, their standardization, flexibility, and redundancy. Redundancy is at the Cubesat level.

Batteries and solar panels

The advanced batteries and solar panels on the ongoing Juno mission at Jupiter are functioning well. These would then be a candidate for the Swarm Mothership. Due to technology advances, solar cells can now be used out to 5 AU. The Mothership will have large arrays, but the individual Cubesats have limited area. They will be deployed fully charged. One area of interest is a deployable fabric solar panel.

At the distance of the asteroid belt, the solar constant (kw/meter^2) is about ¼ of that at Earth's distance from the sun.

Cubesats

The Cubesats will be a mix of 3U and 6U in size (a Unit is 10x10x10 cm), and follow the GSFC-defined PiSat architecture with a Raspberry-Pi flight computer, running NASA/GSFC's CFS/CFE flight software. Different instrumentation will be included on different Cubesats, using common platforms and buses. These will be deployed by the Mothership as required, to observe and collect data on targets of interest. For a lander-Cubesat, a cold gas propulsion system can also be included.

Dispenser/Mothership

The Mothership will be built with standard aerospace

products with a mission heritage. We expect to be able to use the same batteries and solar panels from the Juno mission, since the Mothership will be roughly the same size, defined by the size of the launch vehicle shroud. The X-band transceiver for Juno would be a candidate for the Earth link. The carrier is designed to be modular and adaptable. Standard p-pod Cubesat dispensers are baselined, but the affects of long term storage of the Cubesats in the dispensers in space must be carefully considered. The effects of cold welding during the transit and storage time needs to be evaluated. The PPOD deployment tubes are orientated radially.

Unlike Earth Cubesat missions, the Cubesats going to the Belt can have their own propulsion. The big limiting factor for them is electrical power. They can't carry large solar arrays. Dispersed from the carrier fully charged, they will operate as long as they can. The electronics and software will be optimized to minimize power usage. More advanced solar arrays, fabric-based can serve double duty as a solar sail. This provides both propulsion and power, and can be used with small ion engines.

The Mothership provides cloud services to the swarm. It is a store-and-forward node, and the communications relay to Earth. It provides Swarm control, monitoring, task assignment, and is responsible for Science and engineering data storage.

The Mothership will have a bi-propellant engine for orbit

and cruise adjustments, and a mono-propellant system for attitude control and reaction wheel momentum unloading.

Cubesat Swarm

This section describes a different approach to exploration with collections of smaller co-operating systems that will combine their efforts and work as ad-hoc teams on problems of interest.

Swarm behavior is based on the collective or parallel behavior of homogeneous systems. It is implemented as multiple co-operating software agents. This mimics collective behavior, modeled on biological systems. Examples in nature include migrating birds, schooling fish, and herding sheep. A collective behavior emerges from interactions between members of the swarm and the environment. The resources of the swarm are organized dynamically. Swarm behavior is a group attribute. It relies on continuous co-ordination of behavior. Each member makes stochastic choices. There is a lack of authority in the Swarm, each member making choices based on local conditions, using a global rule set. Example rules are separation, avoid crowding; alignment, do what other swarm members are doing; cohesion, move to the center of mass of the swarm, etc. Swarm activity can be chaotic or orderly. Emergent, un-planned behaviors are seen in implemented systems.

Curtis' paper baselined about 1,000 units. They defined

8-10 types of *Workers,* each with specific capabilities. Units assigned to swarm cohesion and communication they term *Messengers.* There are also Rulers, who function in a managerial role. Cubesats are not specifically mentioned, but the approach is certainly feasible.

They postulated target selection according to mission goals, but also point out that that mission goals change as data is collected at the site. The concept of multiple spacecraft coming together to form virtual instruments is discussed. Here, we can have simultaneous observations from multiple points.

In Swarm systems, the key issues are communication between units, and cooperative behavior. The capability of individual units does not much matter; it is strength in numbers. Self-organizing behavior emerges from decentralized systems that interact both with members of the group, and the environment. Swarm intelligence is an emerging field, and swarm robotics is an active research topic.

The cluster must be agile, able to respond to targets of opportunity, when they arise. Flexibility, and the ability to respond locally to unplanned events will be part of the architecture. A Swarm is an implementation of a multi-agent system, where the agents are implemented in program code.

A constellation of 50, 2U and 3U Cubesats was deployed

in Earth orbit in 2015. Some were released from the ISS, in coordination with a rocket launch. They collected and telemetered data on the lower thermosphere. This was not a swarm per se, but rather 50 units acting on their own, reporting back to their home institutions. Universities around the world participated in the project.

The data came from the region below 85 kilometers, which has enough of an atmosphere to impede spacecraft. The Cubesats collected data as long as they could, as they were reentering the atmosphere. At these altitudes, the rarefied atmosphere can reach temperatures of 2,500 degrees C. It is also a region where the dynamics are controlled by atmospheric tides, themselves controlled by diurnal heating and cooling. The member Cubesats used onboard processing to reduce downlink bandwidth.

Avionics Suite

Both the Mothership and the Cubesats will baseline the GSFC PiSat software and hardware architecture for the flight computers. The Cubesats will use a single unit, and the Mothership will have a 16-unit compute cluster. Non-deployed Cubesats in the Mothership will be able to participate in the clustering, using the Mothership's internal networking infrastructure. The Mothership will be able to power up and attach selected additional units for particularly computational-intensive tasks.

The Raspberry Pi-3 is a very capable processor. An earlier model was tested to operate to 150 k Rad Total Ionizing Dose, with only the loss of several unused I/O interfaces. The major source of radiation at the destination will not be trapped particles, but rather ionizing cosmic rays of galactic origin. These are energetic, but sparse. The cluster computer will be enclosed in the nose of the mothership, with shielding.

A Raspberry Pi-3, requires 3.26 watts of power. It is quad-core, operating at 1.4 GHz. It is a 64-bit machine, with 1 gigabyte of ram, and can achieve 2451 MIPS. It has a dedicated Graphics Processing Unit-based video pipeline that can handle 2D DSP, supported by the Open-GL software library.

Compute cluster of convenience

Using a variation of the Beowulf clustering software and the communications infrastructure of the Mothership, the Cubesats awaiting deployment can be linked into the Mothership's Compute Cluster of Convenience configuration. Each compute node will have the Beowulf software pre-loaded as part of its Linux operating system.

Beowulf was developed at GSFC to provide a low cost solution to linking commodity pc's into a supercomputer. The approach has been applied to clusters of small architectures such as the Raspberry-Pi, that serve as flight computers for Cubesats. Several 64-node Pi clusters have been demonstrated in the Earth environment.

The Beowulf cluster is ideal for sorting and classifying data; an example application for this is the Probabilistic Neural Network (PNN). This algorithm has been used to search for patterns in remotely sensed data. It is computationally intensive, but scales well across compute clusters. It was developed by the Adaptive Scientific Data Processing (ASDP) group at NASA/GSFC. The program is available in Java source code.

The first Beowulf cluster to be flown in space was built from twenty 206-MHz StrongARM (SA1110) processors, and flew on the X-Sat, Singapore's first satellite. The performance was 4,000 MIPS. The cluster drew 25 watts. The satellite was a 100 kg, 80 CM cube. The cluster was used because the satellite collected large amounts of image data (80 GB per day), most of which was not relevant to the mission. An onboard classification algorithm selected which images would be downloaded. For example, cloudy images were discarded, since only land images of Singapore were of interest.

In a cluster, there is always a trade-off of computation, communication, and power draw. This will be monitored and adjusted by the cluster itself, in real-time.

Surface Lander

In the surface exploration scenario, a 6U Cubesat will serve as a local Mothership. It will be able to detach and deploy the lander vehicle. The orbiting Cubesat can provide a "gods-eye" view, to target locations for direct exploration. It will also serve as a communications relay. The main problem will not be mobility, but rather avoiding floating off the surface.

The ground-based unit will implement prox-ops in a more leisurely fashion, in that motion will be two-dimensional and at low speed. The surface rovers will not be retrieved. The local Mothership may be left at the observation site for additional data, or can be redeployed to an additional target. The location of a lander/rover on the surface will be determined by the orbiting or station-keeping Mothership, with imaging. Standard space communications protocols will be used between the lander and the Mothership, via UHF link.

The Cubesat members will collect observational data on their target in multiple bands, and can conduct radio occultation experiments to better categorize the distribution of particles. They can also conduct synchronized simultaneous observations from multiple observation points of features of interest.

NASA is building a "virtual telescope" using two Cubesats. They are imaging the Sun, in a mission called "Cubesat Astronomy by NASA and Yonsei using Virtual Telescope alignment experiment," CanyVAL-X. Two spacecrafts, flying in coordination and aligned with the Sun are being used. The one closest to the sun blocks the solar sphere, allowing the second spacecraft to image the outer regions of the solar atmosphere. The spacecraft are named *Tom* and *Jerry*. Jerry is smaller (1U), and Tom (2U) is between Jerry and the Sun. They both have solar sensors, Tom uses a camera to look at Jerry's laser beacons to keep alignment. The spacecraft are separated by 10 meters. This is an early proof-of-concept mission that will be launched in 2017. The mission costs around $1 million which is a lot for two Cubesats, but a drop in the bucket for a full sized spacecraft.

The Mothership is also an observer, with its own instrument suite. It can search for magnetic fields, and characterize the charged and energetic particle environment.

Communications

Several approaches to communication with spacecraft at large distances from Earth, and examining other planets, have been defined. The Interplanetary Internet implements a Bundle Protocol to address large and variable delays. Normal IP traffic assumes a seamless, end-to-end, available data path, without worrying about the physical mechanism. The Bundle protocol addresses the cases of high probability of errors, and disconnections. This protocol was tested in communication with an Earth orbiting satellite in 2008. It is derived from the mobile data protocols used with cell towers in terrestrial applications. The Disruption Tolerant Network approach is also a good candidate.

As we get farther from Earth, the Cubesat's small antennas, and relatively low power, means we have to get clever with communications. There will be limited bandwidth. This was the case with APL's Horizon's spacecraft at Pluto – It took more than 16 months to transmit all the data back from the encounter.

The deployed Cubesats use the Mothership as their

communication relay, and do not necessarily implement Cubesat-Cubesat direct communications. Cubesat to Mothership will be S-band It does not necessarily need to implement a delay tolerant protocol, since the Cubesats will be "in the vicinity" of the Mothership. Cubesats will exchange two types of data with the Mothership: primary observational data, and secondary metadata which includes position, localization, and timing information.

The Mothership will do onboard processing, as deep as possible, but will always maintain the raw data in a database for a tbd period of time. This depends on the amount of solid state, radiation-protected memory the Mothership can host.

While we would like to transmit the raw observation data, we will be restricted by bandwidth and line-of-sight. Thus, some science data will be processed by the cluster on the mothership, and sent back to Earth via the Deep Space Net when possible. Data processing onboard spacecraft, once a highly controversial issue (and a technical challenge), is now fairly routine.

All data will be kept and transmitted in CCSDS format, stored in a relational database.

An issue in data processing of observational data is the reversibility. Some operations on the data will mean we are no longer able to get back to the raw sensor input. This means, if we have non-reversible operations, and the processing flow is incorrect, we can't access the raw data.

Communication methodology within the Cubesat swarm.

ISL (Inter-satelite link) communication will be achieved on Ultra High Frequency (UHF) links. An inter-satellite communication range of 90 to 100km is viable on UHF within the power output range of 4-5 W. The next challenge is selection of the ideal antenna and communication protocol, keeping in mind the existing power and mobility constraints along with the trade-off between radio power and communication distance. NASA's Nodes (Network & Operation Demonstration Satellite) mission, similar in structure to the Edison Demonstration of Smallsat Networks (EDSN) mission, deployed a satellite swarm of Cubesats from the ISS to test inter-satellite communication capabilities in 2015. A primary UHF radio was used for crosslink communication, and a further UHF beacon radio was used for transmitting real time health information of the satellite. In addition to this, position, navigation, and tracking information complement the primary data load. The Cubesats, and the Mothership will use software defined radio, implemented on the flight computer.

Ops concept

During the cruise phase to the Asteroid Belt, the Cubesats are unpowered. Every day or week (tbd), the units are powered on, one at a time, and checked for functionality.

The onboard database is updated as required. The results are sent back to the control center on Earth. One advantage of the Mothership is, like the Shuttle, payloads can be tested before deployment. Known bad units will be left in place, unpowered, or discarded into Space.

Near the desired target location at the Asteroid belt, the Mothership uses its main engine to enter solar orbit with the solar panels oriented to the Sun, and the high gain antenna pointed to the Earth.
After another system check of itself and the Cubesats, the Mothership deploys a series of Cubesat scouts on a reconnaissance mission, to seek out areas of interest.

The Mothership deploys Cubesats with broad spectral sensing capabilities. Based on their findings, the Mothership may deploy additional Cubesats with specific instrumentation to the area of interest. (For example, an advanced thermal imager to an area of observed thermal activity). The Cubesats are released in the order of necessity. There will only be one Cubesat per dispenser, so blocking is not an issue.

In the surface exploration scenario, the lander necessarily needs a propulsion system for alignment and touchdown. This would be a cold gas system. It would be possible, but more complex, to implement a sample return. This would involve ascent and return to the Mothership, and rendezvous and docking. It is anticipated that the entire Cubesat with its payload would be returned to Earth, which is simpler than a sample hand-off to a return vehicle.

Rad-Hard Software

This is a concept that implements routines that check and self-check, report, and attempt to remediate radiation damage. It is an outgrowth of the testing and self-testing of a computers' functionality, with focus on detection of radiation induced damage. We know, for example, that one of the tell-tales for radiation damage is increasing current draw. At the same time, we monitor other activities and parameters in the system. This partially addresses the problem of operating with non-radiation hardened hardware in a high radiation environment. The baseline RaspberryPi has been radiation tested to 150 kRad, and was operational at that point.

From formal testing results, and key engineering tools, we define likely failure modes, and possible remediation's. Besides self-test, we will have cross-checking of systems. Not everything can be tested by the software, without some additional hardware. First, we use engineering analysis that will help us define the possible hardware and software failure cases, and then define actions and remediation. This is a software FMEA.

failure modes and effects analysis. None of this is new, and the approach is to collect together best practices in the software testing area, develop a library of RHS routines, and get operational experience. Another advantage of the software approach is that we can change it after launch, as more is learned, and conditions change.

Rad Hard software runs in the background on the flight computer, and checks for the signs of pending failure from any known cause. The biggest indicator for radiation damage is an increase in current draw. The mothership cpu cluster monitors and trends current draw across the swarm, and take critical action such as a reboot if it deems necessary. The Rad Hard software will keep tabs on memory by conducting continuous CRC (cyclic redundancy checks). One approach to mitigating damage to semiconductor memory is "scrubbing," where we read and write back each memory locations (being careful not to interfere with ongoing operations). This will be done by a background task that is the lowest priority in the system. Watchdog timers are also useful in getting out of a situation such as a Priority Inversion, or just a radiation-induced bit flip. There will be a pre-defined safe mode for the computer as well. Key state data from just before the fault will be stored. Unused portions of memory can be filled with bit patterns that can be monitored for changes. We must be certain that all of the unused interrupt vectors point to a safe area in the code, so this will be reloaded periodically.

Functions within the RHS include current monitoring as a tell-tale of radiation damage, self-diagnosis suite, spurious interrupt test, memory test(s), checksums over code, data corruption testing, memory scrub, I/O functionality test, peripherals test, stack overflow monitoring, and a watchdog timer. A complete failure modes and effects analysis will be conducted over the flight computer and associated sensors and mechanisms, and this will be used to scope the RHS. The systems will keep and report trending data on the flight electronics. In most cases, the only remediation is a reboot. However, since the units will have identical configurations, the data will be useful to be able to predict pending failures, and to possibly avoid and correct them. This will be used on the Mothership's and on the Cubesat's RaspberryPi-based flight computers. This provides a distributed fault detection and mitigation system, with learning.

We can also choose to implement a small, rad-hard recovery computer, which uses FRAM, which is fairly immune to radiation. The recovery computer would receive heart-beat signals from the cluster members onboard the mothership, and take recovery efforts if they are interrupted. A similar scheme could be used onboard the Cubesats, with little impact on size, weight, power, and cost. This would primarily be used to mitigate latchup.

Onboard databases

Each member of the Swarm is self-documenting. It carries a copy of its Electronic Data Sheet (EDS) description, which can be updated. This defines the system architecture and capabilities, and has both fixed (as-built) and variable entries. The main computer in the Mothership has a copy of all of these, and can get updates by query. The Mothership also has parameters on each unit's state, such as electrical power remaining, temperature, position, etc. One value of the database is, if the Mothership needs a unit with a high resolution imager, it knows what unit that is, and whether it has been deployed or not. If it has been deployed, it will query the unit on its position and health status. Implementing the EDS in a true database has big advantages, since the position of the data item in the database also carries information. It also allows the use of off-the-shelf database tools. The individual Cubesats have a "light-weight" version of the database, while the Mothership has a more sophisticated one. All the schema's are the same. The advantage of a formal database is the structure it imposes on the data.

There are two parts of the tables, representing static and dynamic data. Static Data represents the hardware and software configuration of the swarm unit. These values are not expected to change during the unit's operation. The Dynamic Data table represents the sensors each particular unit has. These values can change, and the last values will be kept. Cubesats will exchange two types of data through their communication channel: primary observational data, along with secondary metadata which includes position and localization information, along with timing information as a part of the EDS during the mission. This approach was prototyped in a previous project.

The Mothership is responsible for aggregating all of the Cubesats' housekeeping and science data, and transmitting it back to Earth. This is also facilitated by the structure imposed by the database. An Open Source version of an SQL database will be implemented. The

EDS documents will be in XML, and the probable database is mySQL, which also has a light version.

Data compression can be implemented onboard the mothership, as well as preliminary data analysis for replanning.

Fly the Control Center

The Mothership is the navigation reference point for the Cubesats. It obtains its position with respect to Earth from observation, and ground tracking. There will be times when the Earth is not visible form the Mothership's position, so it will use extrapolation and local observation. During these periods of occultation, and also periods of long one-way light times, the Mothership assumes local responsibility for the Health and Safety of the Swarm members, and operations of the Swarm. For this, we will implement Control Center functionality within the Mothership. This will take the form of Ball Brother's COSMOS software. This product addresses traditional system test, integration, and flight needs. An additional software module is needed, essentially a virtual Control System Operator. Using defined rules, the Mothership will make decisions concerning the Swarm Members, to the best of its current knowledge. All of this will be documented and downloaded to the Earth-based control center when communications is re-established. An AI capability will be added to Cosmos, in the form of a virtual flight controller agent. Besides the housekeeping functions, we will implement onboard science planning, responsive to on-site conditions, and targets of opportunity.

The Mothership's primary responsibility is continuance of the Mission. To a degree, the Cubesats are considered expendable. During communications black-outs, observations will continue, and the Mothership will dispense explorers according to pre-defined rules, and based on it's best on-scene judgment. It will also continue to collect observation science data, and engineering data related to health and performance across the swarm members.

Wrap-Up

The paper study of a Cubesat mission for Exploration of the Asteroid Belt brought together a series of existing concepts, hardware, and software that would allow implementation of the mission. The core concepts are to use, to the extent possible, Open Source hardware and software, as well as flight-proven high TRL components from other missions. The lessons learned in applying these technology's are applicable in other denseand diverse target areas, such as the moon systems of the Gas Giants, Jupiter and Saturn.

Glossary of Terms

2-D - two dimensional
3-D - three dimensional.
ADAS - Asiago-DLR Asteroid Survey
AIAA – American Institute of Aeronautics and Astronautics.
ALA – asteroid laser ablation.
ANTS – autonomous nano-technology swarm.
AoA – analysis of alternatives.
Aphelion – point in orbit farthest from Sun.
APL – Applied Physics Laboratory of the Johns Hopkins University.
Apogee – point in orbit farthest from the Earth
ARCM – Asteroid Retrieval Crewed Mission.
ARM – asteroid redirect mission.
ARRM – Asteroid Redirect Crewed Mission.
ART – Autonomous Reconfigurable Technology
ARU – Asteroid retrieval and utilization mission (NASA).
ASDP – Adaptive Scientific Data Processing.
ARM – asteroid redirect mission; also a small computer architecture.
Asteroid - minor planet.
Aten asteroid – Earth crossing bodies, 1,300 potentially hazardous.
Atira – class of asteroids within Earth's orbit.
AU – astronomical unit, mean distance from Earth to Sun, 93,000,000 miles.
Avionics – Electronics, including computer, on spacecraft.

Bees - the Biologically Inspired Engineering for Exploration Systems.
Beowulf – computer clustering software, open source, from GSFC.
BP – bundle protocol for delay-tolerant networks.
Carbonaceous chondrite – stony meteorites with magnesium and organic compounds.
CCSDS – Consultive Committee (for) Space Data Systems.
CFE – Core Flight Executive (NASA/GSFC).
CFS – Core Flight Software (NASA/GSFC).
CINEOS - Campo Imperatore Near-Earth Object Survey.
Checksum – error control mechanism for computer memory.
Cislunar – between the Earth and the moon.
CME – Coronal Mass Ejection.
CNSA – China National Space Administration.
CNEOS - Center for Near Earth Object Studies.
Copous – (U. N.) Committee on the Peaceful Uses of Outer Space.
COSMOS – test and operations software from Ball Aerospace.
CPU – central processing unit.
CSA – Canadian Space Agency.
CSS - Catalina Sky Survey.
Cubesat – small, standardized, inexpensive satellite payload.
DART – Double Asteroid Redirection Test
DoD – (U. S.) Department of Defense.
DRO – distance retrograde orbit.
DSP – Digital signal processor.

DTN – delay/disruption tolerant network.
Dwarf planet – a planet not orbiting a star, and not able to clear its neighborhood of other material.
ECAT - Earth Close Approach Table.
EDS – electronic data sheet.
EDSN - Edison Demonstration of Smallsat Networks
ESA – European Space Agency.
FMEA – Failure Modes and Effects Analysis.
FRAM – Ferro-electric RAM, suing magnetic properties to store data,
Gbytes – 10^9 bytes
GHz – giga Hertz - 10^9
GPU – graphics processing unit.
GSFC – Goddard Space Flight Center, Greenbelt, Maryland, NASA Center for unmanned spacecraft near Earth.
HEOMD – (NASA) Human Exploration and Operations Mission Directorate.
HEOSS - High Earth Orbit Space Surveillance.
HST – Hubble space telescope
IAU – International Astronomical Union.
I/O – input/output
ISL – inter-satellite link
IWG – Inter-agency Working Group
JAXA - Japan Aerospace Exploration Agency.
JHU-APL – Johns Hopkins University, Applied Physics Lab
JSC – NASA Johnson Space Center, Texas.
JSOC – U. S. Joint Space Operations Center
JPL – Jet Propulsion Lab.
KT – kilo (10^3) ton.

LAN – local area network.
LARA – ANTS Lander Amorphous Rover Antenna Mission Concept.
LEO – Low Earth orbit.
LINEAR - Lincoln Near-Earth Asteroid Research
LLNL - Lawrence Livermore National Laboratory.
LONEOS - Lowell Observatory Near-Earth-Object Search.
LSST - Large Synoptic Survey Telescope.
MBPL – Minor Body Priority List.
Magnitude - in astronomy, a log scale of brightness.
MEMS – Micro Electronic Mechanical Systems
MASCOT - Mobile Asteroid Surface Scout.
Meteor - glowing meteoroid, or comet, or asteroid passing through the atmosphere.
Meteor Crater – big dent in Arizona. Also known as Barringer Crater.
Meteoroid – rocky or metallic body in space, smaller than 1 meter.
Meteorite – solid debris, enters atmosphere and reaches the surface.
MMOD - micrometeoroid and orbital debris.
MOID - Minimum Orbit Intersection Distance.
Mothership – large vehicle serving multiple smaller vehicles
MPC – Minor Planet Center, Cambridge, Ma.
MPC's – Minor planet circulars, from MPC.
MPO – Minor planets and comets orbit supplement, from MPC.
MPS – Minor Planet Supplement, from MPC.
MSX – mid-course space experiment.

MySQL - an opensource relational database.
NASA – National Aeronautics and Space Administration
NEA – Near Earth Asteroid.
NEAR Shoemaker - Near Earth Asteroid Rendezvous – Shoemaker.
NEAT - Near-Earth Asteroid Tracking.
NIST – National Institute of Standards and Technology.
NEO – Near Earth Object, any significant object within 30 million miles of the home planet.
NEODyS – ESA online database of known NEOs.
NEOP – Near Earth Object Program (JPL).
NEOSSat, Near Earth Object Surveillance Satellite (Canada)
NEOWISE - Near Earth Object WISE (Wide-field Infrared Survey Explorer).
NESS - Near Earth Space Surveillance.
NGO – non-government organizations
NOAA – (U.S.) National Oceanic and Atmospheric Administration.
NODES – Network and Operations Demonstration Satellite
NORAD – North American Air Defense command (USAF).
NSF – (U.S.) National Science Foundation.
NSTC – (U.S.) National Science Technology Council.
Ops - operations
OSIRIS-REx- (Origins, Spectral Interpretation, Resource Identification, Security-Regolith Explore.
OSTP – (U.S.) Office of Science and Technology Policy.
PA&E – Program Analysis and Evaluation.
PAM - – Prospecting Asteroid Mission.

Perihelion – point in orbit closest to the Sun.
Petabyte – 10^{15} bytes.
PHA – potentially hazardous asteroid
PHO – potentially hazardous object.
PI – Principal Investigator.
Pi-SAT – Cubesdat using Raspberry Pi, NASA-GSFC.
PRA – Probabilistic Risk Assessment.
RAD-750 – 32 bit, rad-hard computer.
RHS – rad hard software, and approach for self-check and remediation of radiation damage.
Roche limit – the distance in which an orbiting body will be torn apart by tidal forces of the primary.
SAO – Smithsonian Astrophysics Observatory.
Schema – logical structure of a database.
Semi-major axis – length of the shorter diameter of an ellipse.
SEP – solar electric propulsion
SI – System Internationale (metric).
SMPAG – Space Mission Planning Advisory Group.
Socrates – Satellite Orbit Conjunction Reports Assessing Threatening Encounters in Space.
SQL – structured query language, for database.
SST – space surveillance telescope (DARPA).
Swarm – collection of entities exhibiting collective behavior.
TBD – to be determined.
TECA – Table of Asteroids (Next) Closest Approaches to the Earth.
Thermosphere – layer of the atmosphere about 50 miles above the surface.
TNT – Trinitrotoluene, explosive material.

TRL – technology Readiness Level
U – UNIT, for a cubesat 10cm x 10cm x 10cm
VA – virtual asteroid.
Virtual Satellite – A computer model of a satellite.
XML – Extensible Markup Language, encoding for both human and machine readability.

Bibliography

Alvarez, Jennifer L.; Rice, John R. Samson, Jr., Michael A. Koets. "Increasing the Capability of Cubesat-based Software-Defined Radio Applications,"
 avail: ieeexplore.ieee.org/document/7500847/

Budianu, S.; Engelen, R. T.; Rajan, A. Meijerink; Verhoeven, C. J. M.; Bentum, "The Communication Layer for the OLFAR Satellite Swarm." (2011). avail: doc.utwente.nl/78092/

Challa, Obulapathi N., McNair, Janise "Cubesat Torrent: Torrent-like distributed communications for Cubesat satellite clusters," Military Communications Conference, 2012 , (MILCOM 2012) July 19, 2016. avail: https://www.researchgate.net/.../261237135

Challa, Obulapathi N., McNair, Janise "Distributed Data Storage on Cubesat Clusters," Advances in Computing 2013, (3) 3 pp.36-49. avail: http://article.sapub.org/10.5923.j.ac.20130303.02.html

Clark, P. E.; et al *BEES for ANTS: Space Mission Application for the Autonomous NanoTechnology Swarm,* avail: https://arc.aiaa.org/doi/abs/10.2514/6.2004-6303.

Cudmore, Alan Pi-Sat: A Low Cost Small Satellite and Distributed Mission Test Program, NASA/GSFC Code 582, avail:

https://ntrs.nasa.gov/archive/nasa/casi.ntrs.nasa.gov/20150023353.pdf

Curtis, S. A. et al "Use of Swarm Intelligence in Spacecraft Constellations for the Resource Exploration of the Asteroid Belt," 2003, Third International Workshop on Satellite Constellations and Formation flying, Pisa, Italy, avail.: https://pdfs.semanticscholar.org/

Hall, John "maddog"; Gropp, William *Beowulf Cluster Computing with Linux*, 2003, ISBN-0262692929.

Hinchey, Michael G. ; Rash, James L.; Truszkowski, Walter E.; Rouff, Christopher A., Sterritt, Roy *Autonomous and Autonomic Swarms,* avail: https://ntrs.nasa.gov/search.jsp?R=20050210015-2017-12-20T20:19:24+00:00Z

Jones, Nicola "Tiny 'chipsat' spacecraft set for first flight," Nature, June 2016, Vol 534. avail: www.nature.com/news/tiny-chipsat-spacecraft-set-for-first-flight-1

Kirkpatrick, Brian, et al *Dynamics and Control of Cubesat Orbits for Distributed Space Missions,* Aerospace Corporation, 2015, avail:www.ipam.ucla.edu/wp.../Aerospace-Corp-project-description-FINAL.pdf.

Macdonald, Malcolm "Advances in Solar Sailing," 2014, Springer- Praxis, ISBN-978-3642349065.

Madni, Mohamed Atef; Raad, Raad; Tubbal, Faisal "Inter-Cubesat Communications: Routing Between Cubesat Swarms in a DTN Architecture," presentation, avail: https://iCubesat.org/papers/2015-2/2015-b-2-1.

McLoughlin, Ian; Bretschneider, Timo; Ramesh, Bharath "First Beowulf Cluster in Space," Linux Journal, September 2005, Issue #137, article 8097.

Muri, P., McNair, J. "A Survey of Communication Subsystems for Intersatellite Linked Systems and Cubesat missions," J. Comm., 7 (2012), pp. 290–308.

Nieto, M. M. *The Titus-Bode Law of Planetary Distances: Its History and Theory,* 1972, ASIN-B002M29EWE.

Popescu, Otilia, "Power Budgets for Cubesat Radios to Support Ground Communications and Inter-Satellite Links" Department of Engineering Technology, Old Dominion University, Norfolk, avail: https://ieeexplore.ieee.org/document/7964683/

Rivkin, A. S., et al "Asteroid Studies: a 35-year Forecast, Planetary Science Vision 2050 Workshop, 2017.

Robson, Christopher, "Comparison of Cubesats, Cubesat Swarms and Classical Earth Observation Satellites in LEO, Canadian SmallSat Conference, 2016, avail: https://canadiansmallsatsymposium2016.sched.com/even

t/4bOr/comparison-of-Cubesats-Cubesat-swarms-and-classical-earth-observation-satellites-in-leo

Ruggueri, Marina et al "The Flower Constellation Set and its Possible Applications," Final Report," avail: www.esa.int/act

Saks, N. A.J. Boonstra, R.T. Rajan, M.J. Bentum, F. Beliën, and K. van't Klooster, "DARIS, A Fleet of Passive Formation Flying Small Satellites for Low Frequency Radio Astronomy," The 4S Symposium, Small Satellites Systems and Services, Madeira, Portugal, May-June 2010. avail:
https://ris.utwente.nl/ws/portalfiles/portal/5513988.

Spangelo, et al "JPL's Advanced CubeSat Concepts for Interplanetary Science and Exploration Missions, Cubesat Workshop," 2015, California Institute of Technology, JPL. Avail:
https://digitalcommons.usu.edu/cgi/viewcontent.cgi?article=3313&context=smallsat

Staehle, Robert et al, "Interplanetary Cubesats: Opening the Solar System to a Broad Community at Lower Cost" Cubesat Workshop, 2011, Logan Utah.

Stakem, Patrick H., Kerber, Jonathas, "Rad-hard Software, Cubesat Flight Computer Self-monitoring, Testing, Diagnosis, and Remediation," 2017. available from the author.

Stakem, Patrick H. "Lunar and Planetary Cubesat Missions," Volume 15, Polytech Revista de Tecnologia e Ciência, avail: http://www.polyteck.com.br/revista_online/ed_15.pdf

Stakem, Patrick H.; Da Costa, Rodrigo Santos Valente; Rezende, Aryadne; Ravazzi, Andre, Chandrasenan, Vishnu "A Cubesat-based Alternative for the Juno Mission to Jupiter," 2017, Presented at the Flight Software Conference, FSW-17; avail: http://flightsoftware.jhuapl.edu/files/_site/workshops/2017/

Stakem, Patrick H. *Interplanetary Cubesats*, PRRB Publishing, 2017, ISBN-1520766173.

Stakem, Patrick H. *Cubesat Constellations, Clusters, and Swarms*, PRRB Publishing, 2017, ISBN-1520767544.

Stakem, Patrick H. *Cubesat Engineering*, PRRB Publishing, 2017, ISBN-1520754019.

Stakem, Patrick H. *Cubesat Operations*, PRRB Publishing, 2017, ISBN-152076717X.

Stakem, Patrick H. "Free Software in Space–the NASA Case," invited paper, Software Livre 2002, May 3, 2002, Porto Allegre, Brazil.

Stakem, Patrick H. "Lunar and Planetary Cubesat Missions," March Volume 15, Polytech Revista de

Tecnologia e Ciência, avail: http://www.polyteck.com.br/revista_online/ed_15.pdf

Tan, Ying *GPU-based Parallel Implementation of Swarm Intelligence Algorithms*, 2016, 1st ed, Morgan Kaufmann, ISBN-978-0128093627.

Truszkowski, Walt; Sterrotta, Roy; Rouff, Christopher A.; Hinchey, Michael G.; Rash, James L. "Next generation system and software architectures, Challenges from future NASA exploration missions," 2005, in *Science of Computer Programming* 61 (2006) 48–57.

Truszkowski, Walt, et al *Autonomous and Autonomic Systems: With Applications to NASA Intelligent Spacecraft Operations and Exploration Systems*, Springer, 1st Edition 2009, ISBN-1846282322.

Truszkowski, Walt; Clark, P. E.;, Curtis, S.; Rilee, M. Marr, G. "ANTS: Exploring the Solar System with an Autonomous Nanotechnology Swarm," J. Lunar and Planetary Science XXXIII (2002).

Truszkowski, Walt "Prototype Fault Isolation Expert System for Spacecraft Control," N87-29136, avail: https://ntrs.nasa.gov/search.jsp?R=19870019703,

Truszkowski, Walt, et al, "Ground Autonomy Evolution," 2008, ResearchGate, Publication 251250057.

Violette, Daniel P. "Arduino/Raspberry Pi: Hobbyist Hardware and Radiation Total Dose Degradation, EEE Parts for Small Missions," GSFC, 2014, avail: https://ntrs.nasa.gov/search.jsp?R=20140017620.

Virgili, Bastide et al "Mega-constellations Issues," 41st COSPAR Scientific Assembly, 2016, avail: http://cospar2016.tubitak.gov.tr/en/

Su, WG; Su, FZ; Zhou, CH "Virtual Satellite Construction and Application for Image Classification." 25th International Symposium on Remote Sensing of Environment, Earth and Environmental Science 17 (2014) 012084.

ANTS/SWARM references.

Curtis, S. A.; Rilee, M. L., Clark, P. E., Marr, G. C. "USE OF SWARM INTELLIGENCE IN SPACECRAFT CONSTELLATIONS FOR THE RESOURCE EXPLORATION OF THE ASTEROID BELT," Third International Workshop on Satelllite Constellations and Formation Flying, Pisa, Italy, 24-26, 2003.

Hinchey, Michael G., Rash, James L., Truszkowski, Walter E. "Autonomous and Autonomic Swarms," avail: https://ntrs.nasa.gov/

Johnson, Michael A., Beaman, Robert G., JMica, Joseph A. Truszkowski, Walter F., Rilee, Michael L., Simm, David E. "Nanosat Intelligent Power System Development," avail: https://ntrs.nasa.gov/

Truszkowski, Walt "Prototype Fault Isolation Expert System for Spacecraft Control," NASA/GSFC, 1984.

Truszkowski, Walt et al, "Nanosat Intelligent Power System Development," avail: https://ntrs.nasa.gov/

Truszkowski, Walt, et al "Autonomous and Autonomic Systems, " 2017. avail: https://ntrs.nasa.gov

Truszkowski, Walt et al, A Survey of Formal Methods for Intelligent Swarms, 2004.

Truszkowski, Walt et al, "Next generation system and software architectures Challenges from future NASA exploration missions," 2006, Elsevier,

Clark, P. E., Rilee, M. L. Curtis, S. A., Truszkowski, Walt, Marr, G., Cheung, C. , Rudisill, M. "Bees for Ants: Space Mission Applications for the Autonomous NanoTechnology Swarm," 2004, avail: Researchgate.net

Rilee, Michael L., Stufflebeam, Robert, "ANTS, Autonmous Nanotechnological Swarm,"

Hinchey, Michael G., Sterritt, Roy, Rouff, Chris, Swarms and Swarm Intelligence, IEEE Computer, V 40, Issue 4, April 2007, Consortium on Cognitive Science Instruction.

Clark, P. E., Rilee, M. L., Curtis, S.A. "Exploring with PAM: Prospecting ANTS Missions for Solar System Surveys," 2003, 34th Annual Lunar and Planetary Science Conference, March 17-21, 2003, League City, Texas, abstract no.1493

Mike Hinchey, James L. Rash, Walter Truszkowski, Christopher A. Rouff, Roy Sterritt,:"You Can't Get There from Here! Large Problems and Potential Solutions in Developing New Classes of Complex Computer Systems,". Conquering Complexity, 2012: 159-176.

Hinchey, Michael G., Dai, Yuan-Shun, Rash, James L. Truszkowski, Walt, Madhusoodan, Manish "Bionic

Autonomic nervous system and self-healing for NASA ANTS-like missions."SA 2007: 90-96

Christopher A. Rouff, Michael G. Hinchey, Walter Truszkowski, James L. Rash "Experiences applying formal approaches in the development of swarm-based space exploration systems,". STTT 8(6): 587-603 (2006)

Walt Truszkowski, Christopher A. Rouff, Sidney C. Bailin, Mike Rilee,"Progressive autonomy: a method for gradually introducing autonomy into space missions," ISSE 1(2): 89-99 (2005).

James D. Baldassari, Christopher L. Kopec, Eric S. Leshay, Walt Truszkowski, David Finkel "Autonomic Cluster Management System (ACMS): A Demonstration of Autonomic Principles at Work," ECBS 2005:512-518.

Walt Truszkowski, Michael G. Hinchey, Roy Sterritt: Towards an Autonomic Cluster Management System (ACMS) with Reflex Autonomicity. ICPADS (2) 2005: 478-482.

Christopher A. Rouff,Michael G. Hinchey, James L. Rash, Walter Truszkowski, "Towards a Hybrid Formal Method for Swarm-Based Exploration Missions," SEW 2005: 253-264.

Walt Truszkowski, Mike Hinchey,Rash, James L.. Rouff, Christopher A"NASA's Swarm Missions: The Challenge of Building Autonomous Software," IT Professional 6(5):

47-52 (2004)

Walt Truszkowski, James L. Rash, Christopher A. Rouff, Michael G. Hinchey "Asteroid Exploration with Autonomic Systems." ECBS 2004: 484-489

Christopher A. Rouff, Amy Vanderbilt, Michael G. Hinchey, Walt Truszkowski, James L. Rash:Verification of Emergent Behaviors in Swarm-based Systems. ECBS 2004: 443-448

Christopher A. Rouff, Amy Vanderbilt, Michael G. Hinchey, Walt Truszkowski, James L. Rash: Properties of a Formal Method for Prediction of Emergent Behaviors in Swarm-Based Systems. SEFM 2004: 24-33.

Christopher A. Rouff, Walter Truszkowski, James L. Rash, Michael G. Hinchey, "Formal Approaches to Intelligent Swarms.," SEW 2003:51

from:dblp.unitrier.de

Resources

- Ranging codes: http://www.amsat.org/amsat/articles/g3ruh/123.html

- Planetary Science Vision Workshop - https://www.lpi.usra.edu/V2050/

- Ball Aerospace Cosmos Product,

www.cosmosrb.com

- NASA, https://nssdc.gsfc.nasa.gov/planetary/factsheet/asteroidfact.html

- Small Spacecraft Technology State of the Art, NASA-Ames, NASA/TP2014-216648/REV1, July 2014.

- Core Flight System (CFS) Deployment Guide, Ver. 2.8, 9/30/2010, NASA/GSFC 582-2008-012.

- http://sen.com/news/cubesats-set-for-new-role-as-planetary-explorers

- NASA Systems Engineering Handbook, NASA SP-2007-6105.

- Avail:https://ntrs.nasa.gov/archive/nasa/casi.ntrs.nasa.gov/20080008301.pdf

- wikipedia, various.

If you enjoyed this book, you might also be interested in some of these.

Stakem, Patrick H. *Floating Point Computation*, 2013, PRRB Publishing, ISBN-152021619X.

Stakem, Patrick H. *Architecture of Massively Parallel Microprocessor Systems*, 2011, PRRB Publishing, ISBN-1520250061.

Stakem, Patrick H. *Multicore Computer Architecture,* 2014, PRRB Publishing, ISBN-1520241372.

Stakem, Patrick H. *Personal Robots*, 2014, PRRB Publishing, ISBN-1520216254.

Stakem, Patrick H. *RISC Microprocessors, History and Overview,* 2013, PRRB Publishing, ISBN-1520216289.

Stakem, Patrick H. *Robots and Telerobots in Space Applications*, 2011, PRRB Publishing, ISBN-1520210361.

Stakem, Patrick H. *The Saturn Rocket and the Pegasus Missions, 1965,* 2013, PRRB Publishing, ISBN-1520209916.

Stakem, Patrick H. *Visiting the NASA Centers, and Locations of Historic Rockets & Spacecraft,* 2017, PRRB Publishing, ISBN-1549651205.

Stakem, Patrick H. *Microprocessors in Space*, 2011, PRRB Publishing, ISBN-1520216343.

Stakem, Patrick H. Computer *Virtualization and the Cloud*, 2013, PRRB Publishing, ISBN-152021636X.

Stakem, Patrick H. *What's the Worst That Could Happen? Bad Assumptions, Ignorance, Failures and Screw-ups in Engineering Projects*, 2014, PRRB Publishing, ISBN-1520207166.

Stakem, Patrick H. *Computer Architecture & Programming of the Intel x86 Family*, 2013, PRRB Publishing, ISBN-1520263724.

Stakem, Patrick H. *The Hardware and Software Architecture of the Transputer*, 2011, PRRB Publishing, ISBN-152020681X.

Stakem, Patrick H. *Mainframes, Computing on Big Iron*, 2015, PRRB Publishing, ISBN- 1520216459.

Stakem, Patrick H. *Spacecraft Control Centers*, 2015, PRRB Publishing, ISBN-1520200617.

Stakem, Patrick H. *Embedded in Space,* 2015, PRRB Publishing, ISBN-1520215916.

Stakem, Patrick H. *A Practitioner's Guide to RISC Microprocessor Architecture*, Wiley-Interscience, 1996, ISBN-0471130184.

Stakem, Patrick H. *Cubesat Engineering*, PRRB Publishing, 2017, ISBN-1520754019.

Stakem, Patrick H. *Cubesat Operations*, PRRB Publishing, 2017, ISBN-152076717X.

Stakem, Patrick H. *Interplanetary Cubesats*, PRRB Publishing, 2017, ISBN-1520766173 .

Stakem, Patrick H. Cubesat Constellations, Clusters, and Swarms, Stakem, PRRB Publishing, 2017, ISBN-1520767544.

Stakem, Patrick H. *Graphics Processing Units, an overview*, 2017, PRRB Publishing, ISBN-1520879695.

Stakem, Patrick H. *Intel Embedded and the Arduino-101, 2017,* PRRB Publishing, ISBN-1520879296.

Stakem, Patrick H. *Orbital Debris, the problem and the mitigation*, 2018, PRRB Publishing, ISBN-*1980466483*.

Stakem, Patrick H. *Manufacturing in Space*, 2018, PRRB Publishing, ISBN-1977076041.

Stakem, Patrick H. *NASA's Ships and Planes*, 2018, PRRB Publishing, ISBN-1977076823.

Stakem, Patrick H. *Space Tourism*, 2018, PRRB Publishing, ISBN-1977073506.

Stakem, Patrick H. *STEM – Data Storage and Communications*, 2018, PRRB Publishing, ISBN-1977073115.

Stakem, Patrick H. *In-Space Robotic Repair and Servicing*, 2018, PRRB Publishing, ISBN-1980478236.

Stakem, Patrick H. *Introducing Weather in the pre-K to 12 Curricula, A Resource Guide for Educators*, 2017, PRRB Publishing, ISBN-1980638241.

Stakem, Patrick H. *Introducing Astronomy in the pre-K to 12 Curricula, A Resource Guide for Educators*, 2017, PRRB Publishing, ISBN-198104065X.
Also available in a Brazilian Portuguese edition, ISBN-1983106127.

Stakem, Patrick H. *Deep Space Gateways, the Moon and Beyond*, 2017, PRRB Publishing, ISBN-1973465701.

Stakem, Patrick H. *Exploration of the Gas Giants, Space Missions to Jupiter, Saturn, Uranus, and Neptune*, PRRB Publishing, 2018, ISBN-9781717814500.

Stakem, Patrick H. *Crewed Spacecraft*, 2017, PRRB Publishing, ISBN-1549992406.

Stakem, Patrick H. *Rocketplanes to Space*, 2017, PRRB Publishing, ISBN-1549992589.

Stakem, Patrick H. *Crewed Space Stations,* 2017, PRRB Publishing, ISBN-1549992228.

Stakem, Patrick H. *Enviro-bots for STEM: Using Robotics in the pre-K to 12 Curricula, A Resource Guide for Educators,* 2017, PRRB Publishing, ISBN-1549656619.

Stakem, Patrick H. *STEM-Sat, Using Cubesats in the pre-K to 12 Curricula, A Resource Guide for Educators*, 2017, ISBN-1549656376.

Stakem, Patrick H. *Lunar Orbital Platform-Gateway*, 2018, PRRB Publishing, ISBN-1980498628.

Stakem, Patrick H. *Embedded GPU's*, 2018, PRRB Publishing, ISBN- 1980476497.

Stakem, Patrick H. *Mobile Cloud Robotics*, 2018, PRRB Publishing, ISBN- 1980488088.

Stakem, Patrick H. *Extreme Environment Embedded Systems,* 2017, PRRB Publishing, ISBN-1520215967.

Stakem, Patrick H. *What's the Worst, Volume-2*, 2018, ISBN-1981005579.

Stakem, Patrick H., *Spaceports*, 2018, ISBN-1981022287.

Stakem, Patrick H., *Space Launch Vehicles*, 2018, ISBN-

1983071773.

Stakem, Patrick H. *Mars*, 2018, ISBN-1983116902.

Stakem, Patrick H. *X-86, 40th Anniversary ed*, 2018, ISBN-1983189405.

Stakem, Patrick H. *Lunar Orbital Platform-Gateway*, 2018, PRRB Publishing, ISBN-1980498628.

Stakem, Patrick H. *Space Weather*, 2018, ISBN-1723904023.

Stakem, Patrick H. *STEM-Engineering Process*, 2017, ISBN-1983196517.

Stakem, Patrick H. *Space Telescopes,* 2018, PRRB Publishing, ISBN-1728728568.

Stakem, Patrick H. *Exoplanets*, 2018, PRRB Publishing, ISBN-9781731385055.

Stakem, Patrick H. *Planetary Defense*, 2018, PRRB Publishing, ISBN-9781731001207.

Patrick H. Stakem *Exploration of the Asteroid Belt*, 2018, PRRB Publishing, ISBN-1731049846.

Patrick H. Stakem *Terraforming*, 2018, PRRB Publishing, ISBN-1790308100.

Patrick H. Stakem, *Martian Railroad,* 2019, PRRB Publishing, ISBN-1794488243.

Patrick H. Stakem, *Exoplanets,* 2019, PRRB Publishing, ISBN-1731385056.

Patrick H. Stakem, *Exploiting the Moon,* 2019, PRRB Publishing, ISBN-1091057850.

Patrick H. Stakem, *RISC-V, an Open Source Solution for Space Flight Computers,* 2019, PRRB Publishing, ISBN-1796434388.

Patrick H. Stakem, *Arm in Space*, 2019, PRRB Publishing, ISBN-9781099789137.

Patrick H. Stakem, *Extraterrestrial Life*, 2019, PRRB Publishing, ISBN-978-1072072188.

Patrick H. Stakem, *Space Command*, 2019, PRRB Publishing, ISBN-978-1693005398.

CubeRovers, A Synergy of Technologys, 2020, PRRB Publishing, ISBN-979-8651773138.

Robotic Exploration of the Icy moons of the Gas Giants. 2020, PRRB Publishing, ISBN- 979-8621431006

Hacking Cubesats, 2020, PRRB Publishing, ISBN-979-8623458964.

History & Future of Cubesats, PRRB Publishing, ISBN-979-8649179386.

Hacking Cubesats, Cybersecurity in Space, 2020, PRRB Publishing, ISBN-979-8623458964.

Powerships, Powerbarges, Floating Wind Farms: electricity when and where you need it, 2021, PRRB Publishing, ISBN-979-8716199477.

Hospital Ships, Trains, and Aircraft, 2020, PRRB Publishing, ISBN-979-8642944349.

CubeRovers, a Synergy of Technologys, 2020, ISBN-979-8651773138

Exploration of Lunar & Martian Lava Tubes by Cube-X, ISBN-979-8621435325.

Robotic Exploration of the Icy moons of the Gas Giants, ISBN- 979-8621431006.

History & Future of Cubesats, ISBN-978-1986536356.

Robotic Exploration of the Icy Moons of the Ice Giants, by Swarms of Cubesats, ISBN-979-8621431006.

Swarm Robotics, ISBN-979-8534505948.

Introduction to Electric Power Systems, ISBN-979-

8519208727.

Centros de Control: Operaciones en Satélites del Estándar CubeSat (Spanish Edition), 2021, ISBN-979-8510113068.

Exploration of Venus, 2022, ISBN-979-8484416110.

Patrick H. Stakem, *The Search for Extraterrestial Life,* 2019, PRRB Publishing, ISBN-1072072181.

The Artemis Missions, Return to the Moon, and on to Mars, 2021, ISBN-979-8490532361.

James Webb Space Telescope. A New Era in Astronomy, 2021, ISBN-979-8773857969.

www.ingramcontent.com/pod-product-compliance
Lightning Source LLC
Chambersburg PA
CBHW030505220526
45464CB00006B/2669